THE GEORGIAN GARDEN

THE GEORGIAN GARDEN

An Eighteenth-Century Nurseryman's Catalogue

JOHN KINGSTON GALPINE

Introduction and Commentary on the Catalogue by

JOHN HARVEY

THE DOVECOTE PRESS LTD

First published in 1983 by The Dovecote Press Ltd
Stanbridge, Wimborne, Dorset
ISBN 0 946159 06 8

Designed by Humphrey Stone
Cover illustration by Ian Beck

Printed in an edition of 750 copies by
Biddles Ltd, Guildford, Surrey
and bound by The Newdigate Press Ltd,
Dorking, Surrey

Contents

Acknowledgements

John Harvey would like to record his debt to V.J. Adams Esq and B.G. Cox Esq for making available to him the fruits of their research into the Galpine Family. He also wishes to acknowledge the kind help of Miss M.E. Holmes and the staff of the Dorset County Record Office, and R.N.R. Peers Esq of the Dorset County Museum.

The Dovecote Press would like to thank Mrs S. Tupper of the Blandford Forum Museum Trust for all she has done to make the publication of Galpine's catalogue possible.

Introduction

The history of gardening in England has mostly been told from literary sources and has been concerned with movements of taste and methods of landscape design. Yet by far the greatest contribution made by England, or rather by Britain, to the art of horticulture has been the masterly handling of very large numbers of different kinds of plants. The varying qualities of each plant at different seasons of the year, the shades of green of its leaves or their autumn tints, the harmonies and contrasts of shapes and colours, are the essential stuff of which the English garden has been made. At its most typical it has, besides all this, possessed an indefinable quality of merging into the natural landscape, so that it becomes impossible to distinguish exactly where the forces of the creation have left off and the self-conscious work of man begins.

Archival Sources

As a corrective to an undue reliance upon literary history of gardening as an art form, we have to turn away from the books which tell us what planters thought they were trying to achieve, to contemporary records of what was actually possible. In a few cases these records may be manuscript orders, bills and planting lists with plans, showing precisely what was done in some garden or estate in a given year. The survival of such detailed documentation is rare, and almost entirely confined to great estates. It is impossible to reconstruct from such sources the general state of English horticulture. On the other hand, a great deal can be learned of the limiting factors of horticulture of all kinds, from the study of trade lists and catalogues. These commonly show the complete stock of what was for sale in one place in a certain season. Setting aside the few exceptional cases where money was no object, these lists show exactly what the gardener could do, in a particular year and within reach of that nursery.

The story of nurseries and of their effect upon habits in gardening is, in fact, largely a question of distance and transport. During the Middle Ages, with an extremely limited range of plants and slow and costly land carriage, hardly any plant nurseries existed except near

London and at Oxford and perhaps at some of the greater monasteries. Yet within a couple of generations of the Dissolution of the monasteries there was a fast expanding trade in plants and growing importation of new species from overseas. This fresh trade dealt in two different sorts of plants. From the recently discovered Americas, and from southern Africa, came strange plants of hitherto unknown types, in most cases never before cultivated by man. Unless accompanied by samples of soil and a careful description of habitat and climate, these plants, or their seeds, were unlikely to survive. All too many, in spite of the gardeners' care, perished from the cold and damp of our climate or, perhaps even more often, from being over-coddled and subjected to smoky fumes from charcoal braziers or household fireplaces. Yet it is a tribute to the remarkable adaptability of English gardeners that they did preserve so many of the new harvest of vegetable treasures introduced in this way.

The other sort of plants, those that were brought from the Old World countries of the Mediterranean and the East, had in most cases been long in cultivation. The skill of Persian, Turkish or Arab gardeners had been exercised upon them for centuries, and they had been improved and developed into a multiplicity of hues and shapes, culminating in a whole series of named varieties. Even though the scientific principles of hybridization were unknown, the raising of new varieties from seed was understood as an observed fact, and widely practised. Great fortunes could be made, it was discovered, by playing upon the human desire to form collections. Hence there were tulip manias in Turkey in the sixteenth and eighteenth centuries, in the Netherlands in the seventeenth; and fashions for other flowers: anemones, ranunculus and carnations. Still within living memory is the latest of these intense crazes, that for rhododendrons, which produced the final wave of deliberate exploration for new plants and has now virtually exhausted the reservoir of species. By the middle of the present century all that remained to be done was to produce improvements, by the exercise of patience and ingenuity, upon the available species brought together from every continent and island.

The British Nursery Catalogue

The essential tool of the nurseryman's trade is the list of his stock for sale to the public. The earliest such lists to survive were large broadsheets printed on one side only in several columns. This style continued from the 1670s for about a hundred years, but from the 1720s began to be overtaken by the small book-form list with its own title-page. The first known were issued by Robert Furber, a fashionable nurseryman of Kensington, and comprised one list of forest trees and shrubs, and another of named sorts of fruit-trees. None of the early lists, whether in sheet or book form, printed the prices, which could be changed and quoted at will, without reprinting the catalogue. A few copies with prices added in ink have survived: the first with a complete range of prices for trees and shrubs comes from a firm at Gateshead, Co. Durham, and is of 1754. Ten years later a catalogue with printed prices was issued by John Whittingham of Coventry, but the earliest to include prices of herbaceous plants is apparently that brought out early in 1775 by a Scottish nurseryman, Robert Anderson of Edinburgh, whose firm later became Dicksons Brothers and continued until the beginning of the present century.

From the middle of the eighteenth century the unpriced London catalogues had been arranged by the Latin names for genera, as the influence of Linnaeus's book *Genera Plantarum* of 1737 began the revolution in naming plants, to be effectively completed by his *Species Plantarum* of 1753, now accepted as the foundation of all botanical nomenclature. Even the learned world was slow to accept the work of Linnaeus, and it is remarkable how rapidly the nurserymen in trade discovered the advantages of his simple system. From 1771, when William Malcolm of Kennington in South London issued the first edition of his general catalogue, the more important firms listed plants alphabetically by their Latin names, with the English names in an opposite column. Outside London this form was adopted by John Brunton of Birmingham in 1777 for a book of 92 pages, including many items which he did not keep in stock. In 1782 Brunton's priced catalogue was of trees and shrubs only, and had 22 pages.

In the mean time, from 1775 onwards there had been frequent issues of book-form catalogues giving prices for trees and shrubs, but not for other plants, sent out by several large nurseries in the North of England, especially those of Telford at York and Perfect at Pontefract. The differing types of catalogue show that Britain was effectively divided into two parts, following distinct social attitudes to planting. In London and the whole of the South, where the landowners were immensely wealthy, prices were never stated in print, no doubt because it was a matter of prestige not to count the cost. On the other hand, study of actual bills preserved shows that these southern noblemen and gentry were often very slow to pay, and several nurserymen were forced into bankruptcy by the burden of large accounts left unsettled for years. Further North the cost of planting was carefully calculated before orders were placed, and it was unusual for a bill not to be settled within a year at the outside.

Until the last quarter of the eighteenth century the main emphasis was on planting and improvement of estates and the formation of landscapes on the grand scale. The flower garden, mainly of herbaceous plants, tended to be rather small and stocked with only a few sorts of plant, mainly annuals and biennials sown each year. Bills for seeds bought by a great estate often show the amounts and prices of each kind of vegetable itemized in full, but only a single lump sum for perhaps a dozen packets ('papers') of flower seeds, not named. With the exception of a few collectors of plants of every kind, and of those who planted herbaceous borders and beds in their 'wilderness quarters', there seems to have been a long period when the flower garden was neglected. Only a few big nurserymen in the London area, and one or two in Edinburgh, found it worth while even to list flowering plants other than shrubs.

The Revival of the Herbaceous Garden

In *A History of British Gardening* the late Miles Hadfield made 1780 the end of the great phase of landscape gardening, and 1781 the opening of the picturesque. Naturally there had been a substantial overlap, but this division is very just. A fresh and by far more widely based enthusiasm for gardening and for plants set in, and in 1787 William Curtis founded the *Botanical Magazine*. It was that great periodical, with its beautifully hand-coloured plates of flowering plants both old and new, that was to be principally responsible for a new outlook. It was, all the same, riding on the back of a fashion already in being. So far as surviving evidence goes, this is best proved by the recently discovered catalogue reproduced here. Hitherto the earliest detailed lists of herbaceous plants known from provincial nurseries were one of 1790 from Norwich, and another of Newark-upon-Trent brought out between 1793 and 1795. It might have been thought that the issue of such lists was a consequence of the rapid success of Curtis's *Magazine*. We now know that this was not the case.

Although, like all southern catalogues of the eighteenth century, this list of the stock of John Kingston Galpine of Blandford is unpriced, it is in all other respects amazingly complete. It contains eight pages of hardy trees and shrubs, four pages of Herbaceous plants, four pages devoted to the Greenhouse, and a page and a half of plants for hot-house and dry stove. These are followed by another four pages of named sorts of fruit trees, two pages of seeds of vegetables and salads, and a final three pages of flower seeds and miscellaneous items. Outside the areas of London and Edinburgh, no such comprehensive list of actual stock has up to now come to light.

Detailed consideration of the contents will come later, but several aspects of the historical background must here be mentioned. First of all is the changed emphasis, with the list of trees and shrubs for planting outnumbered by those for the flower garden and greenhouses. Though the hothouse, and even the greenhouse, might remain the preserve of the favoured few, the hardy flowering plants were mostly within the reach of a wide public of owners and occupiers of small gardens. Such a catalogue marks, among other things, the arrival on a grand scale of the amateur gardener. No longer was the cultivation of a large range of plants a matter solely for the head gardener of a great estate, or for an enthusiastic or eccentric nobleman.

What is equally significant is the geographical position of the nursery, at Blandford in the middle of rural Dorset. Just over a hundred miles from London by main road, it was only 15 miles from the then important port of Poole. The cost of heavy carriage overland was a chief factor in early gardening, and it was not until after 1790 that a national system of canals began to remove the limitations set by transport. On the improved turnpike roads, small packages could for the first time be sent down from London rapidly, but larger consignments would arrive by sea, shipped to the nearest port. Even that was still a costly method of buying, and it was this that had led to the establishment of the larger provincial nurseries in the course of the eighteenth century. Once a stock of many different species of plants had been built up at a convenient centre, the practical possibility of adventurous gardening was enormously widened.

The gradual spread of nurseries away from London was mainly conditioned by economic facts of supply and demand, and especially by the disproportionately high cost of transport. The difficulty of transplanting successfully if the journey were slow was an added handicap, and there were frequent complaints of inadequate packing. It is, in fact, surprising that so many trees and plants, carried for long distances, survived and did well. Also surprising is the backwardness of the South-West of England, not only in respect to nurseries, but also in landscaping. Hardly any improved estates are marked in Somerset, Dorset, Devon or Cornwall before the time of Capability Brown (see Hugh Prince, *Parks in England*, Pinhorns, 1967). By 1680 only two substantial nurseries are known to have existed in the West Country, both in Wiltshire, at Westbury and Salisbury. Even by 1760, only Exeter and Crewkerne seem to have been added to the list.

The founding of a large nursery at Blandford forms part of a much wider regional enterprise, which within the last generation of the eighteenth century was to set up plant centres at Bath, Wells, Castle Cary and Taunton in Somerset, at Dorchester, at Honiton, and at Bodmin in Cornwall. By 1800 great firms of almost metropolitan standing had come into being at Bristol, Exeter and Southampton. What is astonishing, and so far revealed only by the chance survival of this Blandford list, is that there were—some twenty years before the end of the century—nurseries at smaller centres carrying a really comprehensive stock. The catalogue of 1782 issued by Galpine of Blandford is, then, a historical document of outstanding importance in horticultural history.

The Blandford Nursery

It has not yet been possible to ascertain the date when the Blandford nursery was founded, but it may have been in about 1775. The credit for the establishment belongs to Francis Kingston junior, nephew of Mr Francis Kingston of Blandford, prominent there by 1743. The younger Francis is mentioned in 1747, but assessed to pay rates of only 4d. on his lands; he was described as 'gardiner' in 1748, but it was not until 1774 that he bought Damory Farm, three years before his death. When he made his will on 1 February 1777 he called himself a nurseryman, and the nursery of Galpine & Son was later said to be theirs as successors to Mr F. Kingston. The younger Francis Kingston left the property to his great-nephew

Francis Kingston Galpine, the elder of the twin sons of his nephew John Kingston Galpine. This John K. Galpine (1741-1786) was the son of Mr George Galpine (died 1752) who was Bailiff of Blandford in 1748, by his wife Elizabeth Kingston, niece of the elder and sister of the younger Francis Kingston of Blandford.

Both the Kingstons and the Galpines were substantial old families, and ranked as gentry. This is of considerable historical interest in that it conforms to an earlier pattern of the founding of nurseries by country gentlemen such as the famous John Rea (died 1677) and Reginald Whitley (*c* 1754-1835), son of the vicar of Merriott near Crewkerne, who moved to London in 1788 to found one of the greatest of the metropolitan nurseries. Francis Kingston Galpine, the elder twin who inherited the Kingston estate of Damory Farm, went up to Oxford, where he matriculated at Oriel College. His brother John Galpine must also have been a man of good education, as he became an Associate of the Linnean Society in 1798 and just before his death (on 6 January 1806) was able to complete his *Synoptical Compend of British Botany*, a valuable reference book which went into three later editions in 1819, 1829 and 1834. John had been apprenticed to his father, John Kingston Galpine, in 1785 when he was 14. It is not clear how the business was carried on after the father's death in April of the next year, but it may have been under John's stepmother Mary until he was of age to take over. The nursery, as Galpine & Son, was being carried on in 1789 when they advertized their new 'Kingston's Portugal Rose' for sale at 7*s*. 6*d*. (37½p.) per plant.

The propagation of this Provence rose was a landmark. Like other new roses of the period, it no doubt derived from a bud-sport observed by the nurseryman on a plant of the older 'Cabbage' rose, like the white Provence or 'Unique' discovered by Daniel Grimwood in 1775 and later propagated for sale, in 1780, at a guinea (£1.05p.) each. The Portugal rose soon became known as the 'Kingston or Blandford' and later as the Blandford, and was illustrated in plate 21 of the *Collection of Roses from Nature* published by Mary Lawrance in 1796-99. It was listed by the botanist Robert Sweet as a variety of 'Rosa provincialis' in 1818 under the name of *Blandford*, or *Portugal*, and in 1827 as the *Blandford*. In his *Encyclopedia of Plants* of 1829 J. C. Loudon called it the 'Blandford or Kingston', but it was listed by George Wheeler of Warminster,

Wilts., in October 1830 as 'Kingston's Portugal', priced at 4*s*. 0*d*. a dozen, the basic price to the trade for the cheapest hardy roses. It is not included in a list of 1000 named varieties with their colours, contributed to the *Floricultural Cabinet* in its second volume of 1834. The Kingston rose had an effective life of only some forty years because, like the other old roses, it was completely overtaken by the new perpetual flowering hybrids based on the China rose introduced in about 1770 but not used for crossing until the 1790s.

Francis Kingston, the presumed founder of the nursery, had died in 1777 and the business passed on to the Galpine family. In this catalogue of five years later there are 40 different sorts of roses listed, but these do not include the new rose, which indeed was being described as new and sold at a correspondingly high price, in 1789. Why, then, was it called Kingston's rather than Galpine's? The only possible conclusion seems to be that it was regarded as a memorial to the founder of the firm, and deliberately used to perpetuate his name. This tends to be borne out by the fact that the names John Kingston were given to the third generation, when John K. Galpine (III) was baptized at Winterbourne Whitechurch on 21 December 1800. His father, J. K. Galpine (II), had married Kitty Clapcott at Blandford on 2 October 1797; Kitty survived for 43 years after her husband's death in 1806 and died in 1849 at the age of 78. The obituaries of the second J. K. Galpine and of his widow both state that he was Paymaster of H.M. 18th Royal Irish Regiment of Foot. He was an Ensign in 1803, Lieutenant in 1805, and apparently held the rank of Captain when he died. This military career of a country gentleman apprenticed to his father's nursery business may seem strange, but the explanation is probably to be found in the renewal of the war against Napoleon in May 1803 after little more than twelve months of peace.

The Galpine family had probably given up the nursery by this time, as the Blandford rate books show that by 1799 one Gould Smith was in occupation of a part only of 'the Great Nursery'. Smith was still a nursery and seedsman, of Bryanston Street, Blandford, when Pigot's directory of 1823 was compiled. In the previous year J. C. Loudon, in his *Encyclopaedia of Gardening*, had given a strangely garbled account in the description of the nurseries of Devonshire: 'The father of Galpine, author of the *Compend of British Plants*, had a nursery at Brenton'. No such place as 'Brenton' exists, and it was probably a confused recollection of Bryanston, Dorset, next to Blandford.

That the nursery was on land leased from the Bryanston estate is made probable by various entries in local records, compared with the tithe map of 1838. The site, owned by the Portman family of Bryanston, but in part leased to Gould Smith after 1799, seems to have occupied not less than ten acres to the north-west of the town, beyond Whitecliff Mill Street and later taken into Bryanston Park. The old nursery was superseded in the later 1820s by a much smaller one of only about two acres behind Dale House at the north end of Salisbury Street. This was run by John Gill, a resident since 1821, and the only substantial nurseryman and seedsman in Blandford for over thirty years. Gill, and his sons after him, maintained the importance of Blandford as a garden centre until 1865, and the tradition was continued still later by Henry Blandford on enlarged grounds.

A

CATALOGUE

OF THE

most Useful & Ornamental

HARDY TREES,

Shrubs, Plants, &c.

ALSO OF THE

Herbaceous Plants, Fruit-Trees, Garden Seeds, Flower Roots, Flowering Shrubs, *&c.*

SOLD BY

JOHN KINGSTON GALPINE

Nursery and Seedsman,

AT BLANDFORD,

DORSET,

Where Catalogues may be had, Price Six-Pence.

BLANDFORD: Printed by S. SIMMONDS, M,DCC,LXXXII.

A CATALOGUE of Hardy Trees, Shrubs, Plants, &c.

	Latin Names.	English Names.
Acer	Rubrum	Scarlet flowering Maple
	Pſeudo Plantanus	Common Sycamore
	Pſeudo Plantanus foliis variegatis	Sycamore with ſtrip'd leaves
	Plantanoides	Norway Maple
Æſculus	Hippo Caſtanum	Horſe Cheſnut
	Pavia	Scarlet flowering ditto
	Flore flavo	Yellow Ditto ditto
Amygdalus	Perſica flore Pleno	Double flowering Peach
	Communis	Common Almond Tree
	Nana	Dwarf Almond
	Pumila	Double dwarf Almond
Andromeda	Polifolia Minor	No Engliſh Name
	Polifolia Major	Ditto
	Caliculata	Ditto
Arbutus	Unedo	Strawberry Tree
	Flore Rubro	Scarlet flowering
Azalea	Viſcoſa	American upright honey ſuckle with white flowers
	Nudiflora	Naked ſtalk'd upright ditto with bluſh colour'd flowers
	Rubra	Red flowering ditto
	Variegata	Red and white ſtrip'd flowering ditto
Baccharis	Halimifolia	Grounſel Tree
Berberis	Vulgaris	Common Berberry
Betula	Alba	Common Birch
	Canadenſis	Canada or paper ditto
Bignonia	Catalpa	Catalpa
	Radicans	Scarlet Trumpet ditto
Buxus	Sempervirens	Common Box Tree
	Myrtifolius	Myrtle leav'd ditto
	Variegatis Argent	Silver ſtrip'd ditto
	Variegatis Aureus	Gold ſtrip d ditto
Calycanthus	Floridus	Carolina Alſpice
Carpinus	Betula	Common Hornbean
	Oſtrya	Hop ditto
	Orientalis	Eaſtern ditto
Celaſtrus	Scandens	Climbing Staff Tree

	Latin Names	English Names
	Bullatus	Round leav'd ditto
Celtis	Auſtralis	European Nettle Tree
Cercis	Siliquaſtrum	Eaſtern Juda Tree
	Canadenſis	American ditto
Chionanthus	Virginicus	Fringe or Snow drop Tree
Clematis	Viticella	Double blue Virgin's bower
	Criſpa	Curl'd flower'd ditto
	Virona	Carolina Ditto
Clethra	Alnifolia	no Engliſh name
Cneorum	Tricoccum	Widow Wail
Colutea	Arboreſcens	Common Bladder ſenna
Coriaria	Myrtifolia	Myrtle leav'd Sumach
Cornus	Florida	Great flowering Dogwood
	Maſcula	Cornelian Cherry ditto
	Sanguinea	Red ſtalk'd Ditto
	Alba	Newfoundland ditto
Coronilla	Emerus	Scorpion Senna
	Coronata	Jointed podded Ditto
Cratæus	Aria	White Beam or Theophraſti
	Acerifolia	Maple leav'd Haw
	Azarolus	Great Azarol
	Oxyacantha	Common Haw
	Ditto flore Pleno	Double flowering Ditto
		Black Virginian Thorn
		Glaſtenbury Ditto
		Lord Iley's Ditto
		Yellow berried Ditto
Cupreſſus	Sempervirens	Upright Cyprus
	Horizontalis	Male ſpreading Ditto
	Diſticha	Deciduous ——ditto
	Thyoides	White Cedar ditto
	Luſitanica	Cedar of Goah
Cytiſus	Laburnum	Baſe tree Trefoil or narrow leav'd Laburnum
	Seſſilifolius	Common ditto, or Cytiſus ſecundus
Daphne	Mezerium Rub.	Mezerium
	Flore Albo	White ditto
	Laureola	Common ditto
	Cneorum	Narrow leav'd moſt beautiful
Erica	Multiflora	Many flowered Heath
Euonimus	Europæus	Common Spindle Tree

	Latin Names	English Names
	Latifolius	Broad leav'd ditto
	Sempervirens	Evergreen Ditto
	Verrucoſus	Warted Ditto
Ephedra	Diſtachya	Shrubby Horſe tail
Fagus	Caſtanea	Spaniſh Cheſnut
	Sylvatica	Common Beach
	Foliis purpureis	Purple leav'd Ditto
Fraxinus	Excelſior	Common Aſh
	Ornus	Flowering Ditto
	Rotundifolia	Manna Ditto
Geniſta	Candicans	Montpelier Broom
	Florida	Spaniſh Dyer's ditto
	Italica	Lucca dyer's ditto
	Sagittalis	Dwarf jointed ditto
	Tinctoria anglica	Dwarf dyer's ditto
Glycine	Fruteſcens	Carolina Kidney-bean Tree
Hedera	Helix	Common Ivy
	Helix variegatis	Strip'd leav'd Ivy
	Quinquefolia	Five leav'd Ivy
Hibiſcus	Flore Purpureo	Purple Althea frutex
	Flore Rubro	Red ——— ditto
	Flore Albo	White ——— Ditto
	Flore variegato	Strip'd flower'd ditto
	Foliis ditto	Strip'd leav'd ———
Hippophæ	Rhamnoides	Sea Buckthorn
Hydrangea	Arborſæmum	
Hypericum	Canarienſe	Canary Tutſon
	Hircinum	Common ſhrubby ditto
Hyſſopus	Officinalis	Common Hyſſop
Jaſſminum	Officinale	White Jaſſamine
	Foliis variegatis	Striped leav'd Ditto
	Fruticans	Common yellow Ditto
	Humile	Italian ——— Ditto
Ilex	Aquifolium	Common Holly
		A collection of ſtriped Holly, 30 ſorts. ———
Juglans	Regia	Common Walnut
	Nigra	Black —— Walnut
Juniperus	Virginiana	Red Virginian Cedar
	Suecica	Swediſh Juniper
	Sabima	Savin
	Foliis variegatis	Striped leav'd Ditto

	Latin Nemes	Engliſh Names
Lavandula	Spica	Spiked Lavender
	Latifolia	Broad leav'd Ditto
	Stæchas	French Ditto
Laurus	Benzoin	Benjamin Tree
Liquſtrum	Vulgare	Privet
Liguidambar	Styracifluum	Sweet Gum
	Aſplenifolium	Fern leaved Gale
Liriodendron	Tulipifera	American Tulip Tree
Lonicera	Caprifolium	Early white Honeyſuckle
	Italicum	Italian ——— Ditto
	Sempervirens	Trumpet ——— Ditto
	Periclymenum	Dutch ——— Ditto
	Quercifolium	Oak leav'd —— Ditto
	Americana	Evergreen —— Ditto
	Balearicum	Minorca Evergreen Ditto
	Nigra	Black berried —— Ditto
	Xyloſteum	Fly ——— Ditto
	Alpigena	Upright Fly —— Ditto
	Cærulea	Blue berried —— Ditto
Lonicera	Symphericarpos	Shrubby St. Peter's Wort
	Diervilla	Yellow Fly ditto
	Glauca	Smooth leav'd ditto
Magnolia	Grandiflora	Great laurel leav'd tulip tree
	Glauca	Leſſer ——— ditto
	Acuminata	Blue flowering —— ditto
	Tripetala	Umbrella Tree —— ditto —
Meſpilus	Germanica	Dutch Medler ———
	Pyracantha	Evergreen Thorn
	Arbutifolia	Arbutus leav'd Service
	Pumila	Dwarf Meſpilus
	Canadenſe	Snowy ditto
	Cotoneaſter	Quince leaved ditto
Morus	Alba	White Mulberry
	Nigra	Black ———Ditto
	Rubra	Red ——— ditto
	Papyrifera	Paper ——— Ditto
Myrica	Gale	Sweet Gale ———
	Cerifera	Candleberry Myrtle———
Paſſiflora	Cærulea	Common Paſſion flower——
Periploca	Græca	Virginian Silk ———
Phillyrea	Media	Olive leav'd Phillyrea
	Anguſtifolia	Narrow leav'd ditto

Latin Names		English Names
	Dentata	Prickly leav'd -- Ditto
	Ilicifolia	Ilex leav'd —— Ditto
Philadelphus	Coronarius	Syringa ————
	Inodorus	Dwarf————Ditto
Phlomis	Fruticoſa	Shrubby Jeruſalem Sage
	Anguſtifolia	Narrow leav'd —— Ditto
Pinus	Sylveſtris	Scotch Pine or Fir
	Pinea	Stone Pine ————
	Strobus	Weymouth Pine ——
	Cedrus	Cedar of Lebanon -—
	Larix	Larch Tree —
	Nigra	Black ———— ditto
	Picea	Silver Fir
	Balſamea	Balm of Gilead Fir
	Canadenſis	White Spruce Fir
	Abies	Norway Spruce ditto
	Rubra	Red Spruce ditto
	Nigra	Black Spruce Fir
Platanus	Orientalis	Eaſtern Plane
	Occidentalis	American Ditto
	Hiſpanica	Spaniſh Ditto
Populus	Alba	White Poplar
	Tremula	Aſpen Tree
	Nigra	Black Poplar
	Lombardienſis	Lombardy Ditto
	Carolinienſis	Carolina ditto
	Balſamifera	Tacamahacca ditto
Prinos	Verticillatus	Winter Berry
	Glaber	Evergreen ditto
Prunus	Padus	Common Bird Cherry
	Ceraſus flore Pleno	Double flowering Cherry
	Virginiana	Carolina Ditto
	Lauro Ceraſus	Common Laurel
	Luſitania	Portugal Laurel
Ptelea	Trifoliata	Shrubby Trefoil
Punica	Granatum	Common Pomegranate
	Flore Pleno	Double flowering ditto
Pyrus	Coronaria	Sweet ſcented Crab
	Sempervirens	Evergreen Ditto
	Baccata	Siberian ditto
Quercus	Phellos	Willow leav'd Oak
	Ilex	Evergreen Ditto

	Latin Names	Engliſh Names
	Suber	Cork Tree
	Rubra	Scarlet Oak
	Robur	Common Engliſh ditto
	Cerris	Turkey Oak
Rhamnus	Catharticus	Common Buckthorn
	Frangula	Berry bearing Alder
	Alaternus	Common Alaternus
	Ditto foliis variegat.	Bloched leav'd Ditto
	Paliurus	Chriſt's Thorn
Rhododendron	Ferrugineum	Alpine Roſe Bay
	Maximum	American —— Ditto
	Ponticum	Purple flowering Ditto
Rhus	Typhinum	Virginian Sumach
	Glabrum	Scarlet Ditto
	Vernix	Verniſh Tree
Robinia	Caragana	Caragana
	Hiſpida	Scarlet Acacia
Roſa	Pimpinellifolia	Red Scotch Roſe
	Ditto foliis variegat.	Striped ditto
	Eglanteria	Common Sweet Brier
	Ditto flore Pleno	Double Ditto
		Semi Double ditto
	Villoſa	Apple bearing ditto
	Centifolia	Dutch 100 leav'd ditto
	Spinoſiſſima	Common Province ditto
	Verſicolor	Roſa Mundi ditto
	Moſchata	Double Muſk Roſe
	Bicolor	Red Auſtrian ditto
	Sempervirens	Evergreen Roſe
	Germanica	Frankfort Ditto
	Alba flore Pleno	Double white ditto
	Lutea	Yellow Roſe
	Hiſpida	Double yellow ditto
	Muſcifera	Moſs Province ditto
	Penſylvanica	Penſylvanian ditto
	Cinnamomea	Cinnamon Roſe
	Pendulina	Great burnet leav'd ditto
	Moſchata ſimplex	Single Muſk ditto
	Theſe are varieties of the afore mentioned ſorts, which have no Specific Names ——	Bluſh 100 leav'd ditto
		Pompone —— ditto
		Roſe of Meux — ditto
		Purple Burgundy ditto

Latin Names		English Names
These are varieties of the afore mentioned sorts, which have no Specific Names.	Marbled Rose do.	Maiden's Blush ditto
	Royal —— do.	Blush Province —— do.
	Damask — do.	Blush —— do.
	York & Lancaster.	Blush Belgic —— do.
	Red —— do.	White Monthly Ditto
	Double Velvet do.	Semi double ditto
	Semi double white	Double Virgin's Rose
	Ditto ——	Single Virgin's Ditto
Rosmarinus	Officinalis	Rosemary
	Foliis variegatis	Striped ditto
Rubus	Idæus	Raspberry
	Fructu Albo	White Ditto
	Fruticosus	Bramble Ditto
	Flore Pleno	Double flowering ditto
	Fructu Albo	White fruited ditto
	Odoratus	Flowering ditto
Ruscus	Aculeatus	Butcher's Broom
	Racemosus	Alexandrian Laurel
Salix	Pentandra	Sweet scented Willow
	Babylonica	Weeping Ditto
Sambucus	Fructu Albo	White berried Elder
	Variegata	Striped leav'd ditto
	Foliis Laciniatis	Parsley leaved ditto
Smilax	Aspera	Rough Bindweed
Solanum	Amara fol variegatis	Striped Solanum
Sorbus	Aucuparia	Mountain Ash
	Demestica	True Service
Spartium	Janceum	Spanish Broom
	Scoparium	Common ditto
	Lusitanicum	White Portugal ditto
Spiræa	Salicifolia	Willow leav'd Spirea
	Hypericifolia	Hypericum frutex
	Crenata	White Spirea
	Opulifolia	Virginian Gelder Rose
Staphylæa	Pinnata	Five leav'd Bladder-nut
	Trifoliata	Three leav'd ditto
Syringa	Vulgaris	Common blue Lilac
	Ditto flore Albo	White Ditto
	Persica	White Persian ditto
	Ditto flore Cærul	Blue —— Ditto
	Laciniata	Cut leav'd blue ditto
Tamarix	Gallica	French Tamarisk

	Latin Names	English Names
	Germanica	German Ditto
Taxus	Baccata	Yew Tree
Thuja	Occidentalis	American Arbor Vitæ
	Orientalis	Chinese Arbor Vitæ
Tilia	Europæa	Common Lime Tree
	Viminibus	Red Twigged ditto
Viburnum	Tinus	Laurestine
	Foliis variegatis	Striped Ditto
	Lucidum	Shining leav'd Laurestine
	Dentatum	Dentated ditto
	Prunifolium	Plumb leaved Viburnum
	Latifolium	Broad leav'd Am. ditto
	Variegatum	Striped leav'd ditto
	Oplelus	Marsh Elder or Gilder Rose
	Flore Pleno	Snow-Ball ditto
Vinca	Major	Great Periwinkle
	Foliis variegatis	Strip'd leav'd ditto
		Double flowering ditto
Vitex	Agnus Castus	Chaste Tree
Yucca	Aloifolia	Adam's Needle
Zanthoxylum	Clava Herculis	Tooth-Ach Tree

HERBACEOUS PLANTS, &c.

Acanthus	Mollis	Smooth leav'd Bears Breech
Achillea	Tomentosa	Yellow Millfoil
Aconitum	Lycoctonum	Yellow Wolfsbane
	Napellus	Early blue ditto
	Pyrenaicum	Fennel leav'd Ditto
	Album	White flowering ditto
Agrostemma	Coronaria flore Pleno	Double Rose Campion
Aletris	Uvaria	No English Name
Alyssum	Saxatile	Yellow Madwort
Anemone	Hepatica	Double Red Hepatica
	Hepatica flore cœruleo pleno	Double blue Ditto
	Hepatica flore cæruleo simplici	Single blue Hepatica

	Latin Names	English Names
	Hepatica flore ſimplici albo	Single White ditto
Anthericum	Liliago	Spiderwort
	Liliaſtrum	Savoy Ditto
Antirrhinum	Majus	Greater Snapdragon
	Foliis variegatis	Striped leav'd ditto
Arum	Triphyllum	Three leav'd Arum
	Ariſarum	Italian Arum
	Foliis variegatis	Striped ditto
Aſphodelus	Luteus	Yellow King's Spear
	Albus	Small White ditto
	Ramoſus	Large White ditto
Aſter	Alpinus	Alpine Starwort
	Amellus	Blue Italian ditto
	Grandiflorus	Great flower'd ditto
	Purpureus	Purple flowered ditto
	Dumoſus	Buſhy ditto
	Ericoides	Heath leav'd ditto
Bellis	Perennis	Double Daiſys. ſix ſorts
Caltha	Paluſtris	Double Marſh Marigold
	Latifolia flore pleno albo	Double white giant Throat-wort
Campanula	Latifolia flore pleno cœruleo	Double blue ditto
	Glomerata	Leſſer ditto
	Decurrens fl. pl. al.	Double white Peach leaved ditto
	Decurrens flore pleno cœruleo	Double blue ditto
	Pyramidalis	Pyramidal ditto
Cheiranthus	Flore pleno	Double bloody Wall-flower
Chelone	Glabra	Red Chelone
	Alba	White ditto
Chryſocoma	Linoſyris	German Goldy Locks
Clematis	Recta	Upright Virgin's Bower
	Integrifolia	Entire leav'd ditto
Convallaria	Majalis	Common Lily of the Valley
	Majalis flore pleno	Double ditto
	Majalis flore rubro	Red flowered ditto
	Polygonatum	Upright ditto
	Latifolia	Broad leaved ditto
Coreopſis	Verticillata	Larkſpur leaved Sunflower

	Latin Names	English Names
	Tripteris	Three leav'd ditto
	Lanceolata	Long leav'd ditto
	Alternifolia	Alternate leaved ditto
Delphinium	Grandiflorum	Great flowering Larkſpur
	Elatum	Siberian Bee ditto
Dianthus	Barbatus flore pleno	Double Dutch Sweet Will'm
Dictanus	Albus	White Dittany or Fraxinella
	Flore rubro	Red ditto
Digitalis	Ferruginea	Iron colour'd Foxglove (*ful*
Dodecathean	Meadia	American Cowſlip *moſt beauti-*
Epilobium	Anguſtifolium	Red French Willow
Eryngium	Maritimum	Common Sea Holly
Fumaria	Radix Cava	Hollow rooted Fumatory
	Sempervirens	Evergreen ditto
Gentiana	Purpurea	Purple Gentian
Geranium	Phæun	Black Cranesbill
	Striatum	Streaked ditto
	Maculatum	Spotted ditto
	Sanguineum	Bloody ditto
Gnaphalium	Plantaginifolium	Plantain leaved Everlaſting
Helianthus	Multiflorus	Many flower'd Sunflower
Helleborus	Niger	Chriſtmas Roſe
	Viridis	Green flowering Hellebore
	Fœtidus	Bear's-foot Hellebore
	Trifolius	Three leav'd Hellebore
Heſperis	Flore pleno albo	Double white Rocket
	Flore pleno rubro	Double red ditto
Iberis	Sempervirens	Evergreen Candy-tuft
Iris	Suſiana	Calcedonian Flower de Luce
	Florentina	White Florentine Ditto
	Variegata	Variegated Ditto
	Pumila	Dwarf Ditto
	Graminea	Graſs leav'd Ditto
	Tuberoſa	Snake's-head ditto
Lamium	Orvala	Great purple Archangel
Lobelia	Cardinalis	Scarlet Cardinal Flower
	Siphilitica	Blue Ditto
Lupinus	Perinnis	Perennial Lupine
Lychnis	Chalcidonica fl. pl.	Double Scarlet Lychnis
	Flos cuculi fl. pleno	Double ragged Robin
	Viſcaria flore pleno	Double Catchfly
Lyſimachia	Vulgaris	Common yellow Looſe-ſtrife

	Latin Names	English Names
	Tenella	Purple ditto
Matricaria	Flore pleno	Double Feverfew
Monarda	Fiſtuloſa	Scarlet flowering Oſwego tea
Oenothera	Anguſtifolia	Narrow leav'd Tree Primroſe with a red ſtalk
	Pumila	Dwarf ditto
Pæonia	Fœmina	Female Pæony
	Ditto fl. pl. purp.	Ditto with Double purple Flowers
Phlox	Carolina	Carolina Lychnidea
	Alba	White ditto
Plantago	Roſea	Roſe Plantain
Potentilla	Erecta	Cinquefoil
Pulmonaria	Virginica	Virginian Lungwort
Ranunculus	Ficaria Flore pleno	Double Pilewort
	Aconitifolius fl. pl.	Double white Mountain Ranunculus
	Acris flore Pleno	Double yellow upright ditto
Rheum	Rhaponticum	Rhapontic Rhubarb
	Palmatum	True Ditto
Rudbeckia	Purpurea	Purple flower'd Ditto
	Atropurpurea	Dark purple Sunflower
Rumex	Sanguineas	Bloody Dock
Sambucus	Ebulus	Dwarf Elder
Saxifraga	Cotyledon	Pyramidal Saxifrage
	Granulata fl. pl.	Double white Saxifrage
	Craſſifolia	Siberian Saxifrage
Serratula	Squarroſa	Squarroſe Saw-wort
Solidago	Sempervirens	Evergreen Golden Rod
	Alpina	Dwarf alpine Ditto
Spiræa	Aruncus	Goat's beard Spiræa
	Filipendula fl. pl.	Double Dropwort
	Lobata	Purple flowering ditto
Stipa	Pennata	Feather Graſs
Tradeſcantia	Virginiana	Virginian blue Spiderwort
	Ditto flore albo	White ditto
Trollius	Europæus	Globe Ranunculus
Valeriana	Rubra	Red Garden Valerian
	Flore albo	White Garden ditto
Veratrum	Album	White Hellebore
	Nigrum	Black ditto
Veronica	Virginica	Virginian Speedwell

Latin Names.		English Names.
	Ditto Incarnata	Flesh coloured ditto
	Flore albo	White Speedwell
	Longifolia	Long leav'd ditto
	Spicata	Spiked ditto
	Montana	Mountain ditto
Viola	Odorata fl. pl.	Double Purple Violet
	Ditto fl. pl. albo	Double white ditto

GREEN-HOUSE PLANTS.

Agave	Americana	Large American Aloe
	Ditto fol. variegat.	Variegated Ditto
Aletris	Uvaria	Iris Uvaria
Aloe	Humilis	Dwarf Hedgehog Ditto
	Variegata	Partridge breast ditto
	Disticha	Tongued ditto
	Verucosa	Pearled tongue ditto
	Retusa	Cushion ditto
	Margaritifera	Great Pearl ditto
Alstromœria	Pelegrina	No English Name, most beautiful
Arctotis	Aspera	no English Name
Ascyrum	Crux Andræa	St. Andrew's Cross
Cactus	Moniliformis	Indian Fig
Calendula	Fruticosa	Shrubby Cape Marigold
Canaria	Campanula	Canary Bell-flower
Cassine	Paragua	South Sea Tree
	Maurocenia	Hottentot Cherry
	Ditto Minor	Lesser ditto
	Capensis	Cape Phyllirea
Celastrus	Buxifolius	Box-leav'd Staff Tree
	Pyracanthus	Pyracantha leav'd ditto
Ceratonia	Siliqua	Carob or St. John's Bread
Cistus	Pilosus	Hairy Cistus
	Populifolius	Poplar leav'd ditto
	Creticus	Cretan ditto
	Halimifolius	Atriplex-leav'd ditto
Citrus	Medica	Lime Tree

	Latin Names	English Names
	Limonium	Lemon Tree
	Ditto variegatis	ſtrip'd Ditto
	Aurantium	Seville Orange ditto
	Ditto Sinenſis	China Orange Ditto
	Decumana	Shadock
	Foliis variegatis	Striped leav'd Orange
Clutia	Alaternoides	No Engliſh Name
Colutea	Fruteſcens	Scarlet Bladder Senna
Convolvolus	Cneorum	Silver leav'd Convolvolus
Coronilla	Glauca	Seagreen Colutea
Craſſula	Coccinea	Scarlet Orpine
	Cultrata	——
	Tetagrona	Square leav'd Orpine
Cyclamen	Indicum	Perſian Sow-Bread
Digitalis	Canarienſis	Canary Foxglove
Dioſma	Oppoſitiſolia	African Spiræa with oppoſite leaves ——
	Rubra	Red Ditto
	Ericoides	Heath-leaved Ditto
Dracocephalum	Canarienſe	Balm of Gilead
Erica	Multiflora	Many flowered Heath
	Triflora	Three flowered Ditto
	Mediterranea	Mediterranean ditto
Euphorbia	Caput Meduſæ	Heduſa's Head
Geranium	Inquinans	Scarlet Crane's-bill
	Papilionaceum	Butterfly ditto
	Cucullatum	Hollow leav'd Ditto
	Peltatum	Ivy-leav'd Ditto
	Zonale	Horſe-ſhoe ditto
	Variegatum Argent.	Silver ſtrip'd Ditto
	Ditto Aureum	Gold Striped ditto
	Vitifolium	Balm ſcented ditto
	Capitatum	Roſe ſcented ditto
	Odoratiſſimum	Sweet ſcented ditto
	Coriandrifolium	Coriander leav'd Ditto
	Gibboſum	Gouty Ditto
	Groſſularioides	Gooſeberry leaved ditto
	Triſte	Night ſmelling Ditto
Gnaphalium	Orientale	Eaſtern Everlaſting
Hermannia	Althæifolia	No Engliſh Name
	Alnifolia	Ditto
	Groſſularifolia	Ditto

	Latin Names	English Names
Hypericum	Balearicum	Warted St. John's Wort
	Monogynum	China ditto
Jasminum	Grandiflorum	Spanish Jasmine
	Azoricum	Azorian ditto
	Odoratissimum	Yellow Indian Ditto
Iberis	Semperflorens.	Shrubby Candy-tuft
	Ditto foliis variegat.	striped leav'd ditto
Juniperis	Bermudiana	Bermudas Cedar
Justicia	Adhatoda	Malabar Nut
Ixia	Crocata	No English Name
Laurus	Camphora	Camphire Tree
Lycium	Afrum	Box Thorn
Medicago	Arborea	Shrubby Moon Trefoil
Melia	Azederach	Bead Tree
Melianthus	Major	Great Honey Flower
Mesembryanthemum	Noctiflorum	Night flowering Marigold
	Splendens	Shining flowered ditto
	Umbellatum	Umbellated Ditto
	Calamiforme	Quill leav'd Ditto
	Bellidiflorum	Daisy flowered Ditto
	Deltoides	Delta shaped Ditto
	Barbatum	Bearded ditto
	Humile	Dwarf bearded ditto
	Hispidum	Prickley stalk'd ditto
	Striatum	Striped flowered ditto
	Villosum	Hairy ditto
	Emarginatum	Notch'd flowered ditto
	Uncinatum	Hooked leav'd ditto
	Spinosum	Prickly ditto
	Tenuifolium	Slender leav'd ditto
	Stipulaceum	Upright shrubby ditto
	Crassifolium	Creeping ditto
	Falcatum	Crooked leav'd Ditto
	Glomeratum	Crowded ditto
	Filamentosum	Thready ditto
	Acinaciforme	Scimetar leav'd ditto
	Forficatum	Forked ditto
	Bicolorum	Two colour'd ditto
	Serratum	Sawed Ditto
	Micans	Glittering Ditto
	Glaucum	Sea-green Ditto
	Corniculatum	Horned ditto

	Latin Names	English Names
	Expanſum	Broad leav'd Ditto
	Ringens Caninum	Dogs chop ditto
	Ringens felinum	Cats chop Ditto
	Dolabriforme	Hatchet leav'd ditto
	Difforme	Deformed ditto
	Linguiforme	Tongue ſhaped ditto
	Ditto Pugioniforme	Dagger-leav'd ditto
	Aureum	Orange flower'd ditto
Myrica	Quercifolia	Oak-leav'd Candleberry Myrtle
	Cordifolia	Round leav'd ditto
Myrſine	Africana	No Engliſh Name
Myrtis	Communis	Common narrow leav'd Myrtle
	Romana	Broad leav'd ditto
	Tarentina	Box leav'd ditto
	Bœtica	Orange leav'd Ditto
	Belgica	Double flowering Ditto
	Mucronata	Nutmeg Ditto
	Roſmarinifolia	Roſemary leav'd Ditto
	Foliis reflexis	Thyme leav'd ditto
	Fol. variegat. argent.	Silver Striped Ditto
	Fol. variegat. aur.	Gold Striped Ditto
Nerium	Oleander Rubrum	Red Roſebay
	Oleander Album	White ditto
	Antidiſentericum	Double ditto
	Zeylonicum	Leſſer ſweet ditto
Olea	Europæa	Common Olive
	Capenſis	Cape ditto
Origanum	Dictamnus	Dittany of Crete
	Sipyleum	Dittany of Mount Sipylus
Paſſerina	Filiformis	Ethiopian Sparrow-wort
Philadelphus	Aromaticus	No Engliſh name
Phyllica	Ericoides	Baſtard African Heath
Polygala	Myrtifolia	African Milkwort
Rhus	Tomentoſum	Hairy three leav'd Sumach
	Anguſtifolium	Narrow leav'd ditto
	Lucidum	Shining leav'd Ditto
Roſa	Indica	China Roſe
Roſmarinus	Officin. ſol. variegat.	Striped Roſemary
Royena	Lucida	African Bladder Nut with ſhining leaves ——— }

	Latin Names	Engliſh Names
Ruſcus	Hypophyllum	Alexandrian Laurel
	Androgynus	True Ditto
Selago	Spuria	No Engliſh Name
Sempervivum.	Arboreum	Tree Sedum
	Foliis variegatis	Variegated Ditto
Solanum	Pſeudocapſicum	Winter Cherry
Tarconanthus	Camphoratus	African Fleabane
Teucrium	Fruticans	Tree Germander
	Latifolium	Broad leav'd Ditto
	Marum	Cat Thyme
Trachelium	Cœruleum	Blue Throatwort
Viburnum	Glabrum	Smooth Wayfaring Tree
Yucca	Aloiſolia	Aloe leav'd Adam's Needle

HOT-HOUSE and DRY STOVE PLANTS.

Aletris	Capenſis	No Engliſh Name
Amaryllis	Atamaſco	Atamaſco Lily
	Formoſiſſima	Jacobæan ditto
	Criſpa	
	Regina	Mexican ditto
	Belladonna	Belladonian ditto
	Longifolia	Long leav'd ditto
Anthericum	Aloides	Aloe leaved Spider Wort
	Aſphodeloides	Onion leaved ditto fruticans
Apocynum	Fruteſcens	Shrubby Dogſbane
Bromelia	Ananas	Mountſerrat Pine
	Theſe have no Spe-	Black Antigua Ditto
	cific names———	Queen ditto
Cacalia	Kleinia	No Engliſh Name
	Ficoides Minor	Ditto
	Ditto Major	Ditto
Cactus	Tetragonus	Four angled Torch Thiſtle
	Pentagonus	Five angled ditto
	Fagelliformis	Creeping Cereus
	Opuntia	Oval leav'd Indian Fig
	Mammillaris	Leſſer Melon Thiſtle

	Latin Names	English Names
Ceſtrum	Nocturnum	No Engliſh Name
	Diurnum	Ditto
	Montanum	ditto
Chironia	Fruteſcens	ditto, moſt beautiful
Cotyledon	Lacineata	Cut leaved Navelwort
Craſſula	Nova	No Engliſh Name
	Cultrata	ditto
	Portulacaria	ditto ———————
Euphorbia	Neriifolia	Oleander leav'd ditto
Gardenia	Florida flore Pleno	Double Cape Jaſmine
Geranium	Lobatum	Vine leav'd Crane's-bill
	Triſte	Anemony leaved ditto
	Fulgidum	Celandine leav'd ditto
Gladiolus	Spicatus	Spiked Cornflag
Hæmanthus	Coccineus	Blood Flower
	Puniceus	Spotted ſtalk ditto.
Heliotropium	Peruvianum	Sweet Turnſole
Hibiſcus	Roſa Sinenſis.	Double China Roſe
	Mutabilis	Changeable ditto
	Abelmoſchus	Muſk ditto
Juſticia	Picta	Painted Malabar Nut
Ixia	Crocata	No Engliſh Name
Lantana	Involucrata	ditto.
	Camara	ditto ———————
	Aculeata	ditto
	Odorata	ditto ———————
Martynia	Perennis	No Engliſh Name
Mimoſa	Nilotica	Egyptian ditto
	Pudica	Senſitive Plant
Oxalis	Incarnata	Pink colour'd Wood Sorrel
Paſſiflora	Murucuja	Moon ſhaped ditto
	Holoſericea	Silk leav'd ditto
Plumbago	Zeylanica	Ceylon Leadwort
Rivinia	Humilis	No Engliſh Name
Stapalia	Variegata	ditto ———————
	Hirſuta	ditto
Turnera	Ulmifolia	ditto
Vinca	Roſea	Madagaſcar Periwincle
	Alba	White ditto — ———

FRUIT TREES.

APPLES:

Table Fruit.

DENNETT'S Nonpareil
Dutch Codlin
Englifh ditto
June Eating
Loan's Pearmain
Quince Apple
Golden Rennete
Aromatic Pippin
Hertfordfhire, or Winter Pearmain
Kentifh Pippin
Holland ditto
Monftrous Rennete
Lemon Pippin
Royal leather coat Ruffet for Baking
Wheeler's Ruffet for ditto
Pile's ditto for ditto
Nonpareil
Golden Pippin
King Apple
Julyflower
Dufan
Bayard
Summer Streaking
Marigold
Famagufta
Ruffet Harvey
Kernel Apple
Pomeroy
London Pippin
Nonfuch
Ifle of Wight Pippin
Golden Ruffet
Canary ditto
Orange ditto
Sheepfnout

CYDER FRUIT.

BLANDERLING
Flood Hatch
Jerfey
Ten Bufhels
True Briton
Sour Redftreak
Snaiden's ditto
Walk's fweet Apple
Midyate
Cadbury Pouncer
Cogagee
Stire
Ibberton Redftreak
Piddle ditto
Herefordfhire ditto
Royfe's Buckland
Sour bitter fcauld

PEARS.

LITTLE Musk Supreme
Little Muscat
Cuisse Madam
Windsor
Jargonelle
Blanquet
Avorat
Summer Boncretien
Orange Bergamot
Summer ditto
Autumn ditto
Hambden's ditto
Jansel's ditto
Brocus's ditto
Grey Beuree
Beuree Rouge or brown Beuree
La Marquis
Messieur Jean
Swan Egg
Crasane
Colmar's
St. Michael
Virgouleuse
St. Germain
Spanish Boncretien for Baking
Parkinson's Warden, or black Pear of Worcester for Ditto
Winter Boncretien
Cadilliac for Baking
Union, or Uvedale's St. Germain for Ditto
Chamontelle
Double flowering
Easter Bergamot
Green Juneating
Catharine
Stainer's Early
Dutch Warden

PLUMBS.

ORLEANS
Green Gage
Red Fotheringham
Blue Perdrigon
La Royal
Semiano
Imperial, or Red Magnum bonum
Jeane Hative
Queen Mother
Imparatrice
White Perdrigon
Violet Primordian
Early blue Violet
Drab d'or
Maitre Claude
Reine Claude
Violet Perdrigon
Morocco
Red Perdrigon
Rushia Corboon
Apricot Plumb
Royal Dauphin
White Gage
Double flowering
La Mirable
Cherry Plumb
White Magnum bonum
Verdock

APRICOTS.

PEACH Apricot
Moor Park
Bruſſels
Breda
Algeir
Blotch'd leav'd
Turkey
Early Maſculine
French Maſculine
Roman
Tranſparent

CHERRIES.

BLACK Heart
White heart
Ox heart
Bleeding Ditto
Black Spaniſh Ditto
Amber Ditto
May Duke
Early Duke
Holman Ditto
Kentiſh, or Cluſter
Morrello
Carnation
Carroon

PEACHES.

ROYAL George Peach
Golander
Portugal
Incomparable
White Magdalen
Red Magdalen
French Royal
Port Royal
Early Newington
Early Ann
Red Nutmeg
Early Admirable
Late Admirable
Catharine
Vanguard
Nobleſt
Orange
Duke
Montaubon
Sion
Smith's Newington
Mignon
Matchleſs
Dragon
Double flowering
Millet's Mignon
Nivette
Pavey Royal
Monſtrous Pavey
Tetton de Venus
Bourdine
Bloody

PEACH TREES Continued.

Belle Chevereuſe
Double Swalch
Chancellor
Mr. Portman's Early
Bourdeaux
Buckingham
Rumbullion
Keſhorton
Double Hemſkirk

NECTARINES.

FAIRCHILD Early
Elruge
Newington
Red Róman
Brugnon
Murry
Temple
Peterborough, or late Green
Golden
Tawney
Violet Muſk
Yellow Italian
Early Pavey
Scarlet
Aromatic
Clearmount
White

GRAPES.

BURGUNDY Miller, or Black cluſter ———
Black ſweet water
White ſweet water
Royal Muſkadine for forcing
St. Peter's for ditto
Black Frontiniac
White ditto for forcing
Red Hamborough for ditto
Claret Grape
Parſly leaved
Raiſin for forcing
Red or Griſley Frontiniac

FIGS.

BROWN Iſchia
Black Genoa
Early white
Large ditto
Black Iſchia
Brunſwick, or Hanover
Common blue, or purple

GOOSBERRIES
CURRANTS
RASPBERRIES
STRAWBERRIES, &c.

SEEDS of Roots.

ONIONS.

Deptford	Strafburg	Silver Skin
White Spanifh	Portugal	Blood Red

TIME of Sowing, February till May for Winter Ufe, and till September for young Onions.

Welch, | July till September.

LEEKS.

LONDON
Flanders } February till June.

CARROTS,

Orange	Early Horn	Swelling Parfnip

February till May, and in July, earlier on warm Borders.

TURNIPS.

Early Dutch	White round	Red round	French
Stone	Green round	Yellow	Tankard

March or April for Early, and all the Summer Months, but July and Auguft moft ufual Time.

RADISHES.

Early Dwarf Red
Early ditto Salmon ——— } January till October
London fhort Top

Black Spanifh
White Spanifh ——— } For Winter the middle of July.
Turnip

Skirret	Scorzonera	Salfafy

——— February till May. ———

Shallots	Garlick	Rocambole

February or March, or in September.

SALLAD SEEDS.

LETTUCES.

White Cofs	Black Cofs	Silefia	Roman Cabbage
Green Cofs	Egyptian green	Brown Dutch	Common ditto
Spotted Cofs	ditto for Stoves.	Capuchin	Ball or Button
Bloody or red do.	Imperial	Grand Admiral	Hardy Green

February till September, earlier on Hot Beds or warm Borders.

Corn Sallad. Feb. and Aug. or Sep.

En-

ENDIVE.

White Curl'd	Green Curl'd	Batavia

Middle of May to Sep. The Batavia best for the Winter

PARSLEY.

Common	Curl'd	Hamburgh

February till September

SPINAGE.

Round	Prickly	French

CRESSES.

Common	Curl'd	Broad Leav'd

White Mustard	Rape	Sorrel
Radish	Chervil	Burnet

Feb. to September, earlier on Hot-Beds.

PURSLANE.

Green,	Golden

Apr. to June, earlier on Hot-Beds

CELERY.

Large Italian	Celeriac	Solid,

Mar. till June

BEET.

White, Red, Green, — March till May, and in July.

CHARDON	FINOCHIA	ASPARAGUS

May, middle of August, Feb. or March.

CUCUMBERS.

Early Prickley	Short Prickly	Green Turkey
Long Prickly	White Turkey	

January till May.

Mellon of Sorts. — Jan. or February and March.

Pompions. Gourds of Sorts. — April or May

SEEDS OF GREENS.

CABBAGES.

Cauliflower	Early Sugarloaf	Late long Sided
Wrench's Ear-Dwarf	Red Dutch	Turnip
Early York	Russia	Scotch for Cattle
Early Battersea	Large flat Winter	

July or August for early, February till May for late

BROCOLI.

White, Purple, and Green; May till July.

SAVOY.

Green, and Yellow. — — —
Colewort, Curled Brown Cole. Green Curled
Scotch do. Sea ditto, and Chou de Milan; — } March till July.

SWEET AND POT-HERB SEEDS.

Thyme	Winter Savory	Rhue	Borrage
Sweet Marjoram	Rosemary	Hyssop	Clary
Pot Marjoram	Baum	Lavender	Sweet Bazil
Summer Savory	Sage	Marigold	————

March till May.

PHYSICAL SEEDS.

Carduus	Fennel	Gromwell	Caraway
Scurvy-grass	Tobacco	Fenugreek	Coriander
Angelica	Plantain	Cummin	Oculus Christi
Lovage	Nettle	Rhubarb	————
Dill	Henbane	Anniseed	————

March or April.

TENDER ANNUAL FLOWER SEEDS.

Prussian WhiteWall	Diamond Ficoides	African Marygold
Purple Wall	Stramonium	Cape Marygold
Scarlet ditto	Capsicum	China After, Sorts
Minionet	Egg Plant	Convolvolus Major
Coxcomb	Love Apple	Convolvolus Minor
Tricolor	Marvel of Peru	Sweet Sultan, Red and Yellow
Globe Amaranthus	Balsam	
Sensitive Plant	Indian Pink	Sweet Scabious
Humble Plant	French Marygold	Portugal Bush Bazil
Chrysanthemum, white and yellow	Zenia Peruvia, red and yellow	Large Spotted China Pink

February till May, on a moderate Hot-bed, and transplanted into Pots or Beds; or most of them may be sown on a warm Border, in a rich Earth, in April, or from March till May, as the Season admits.

HARDY ANNUAL FLOWER SEEDS.

Nolana	Belvedere	Larkspur, upright and branched
Heart's Ease	Nigella Romana	
Venus's Looking glass	Flos Adonis	Carnatian and Dwarf Poppy
Venus's Navel Wort	Cyanus or Corn Bottles	
Catchfly		Hawkweed
Virginia Stock	Ten-week Stock, 4	Love lies Bleeding
Princes Feather	sorts ————	Lavatera

Horn Flower	Candytuft, ſorts	Mountain Lychnis
China Mallow	Perſicaria	Dwarf & tall Sun-flowers
Naſturtion	Snail flower	
Caterpillars	Hedgehogs	Lupins of various ſorts
Sweet-ſcented Peas	Tangier Peas	Wing Peas
Sweet Trefoil	Ketmia	Curled Mallow ——

March till May, in Beds or Borders of good Earth; or ſome of them may be ſown in Autumn, to have early flowers in the Spring.

BIENNIAL FLOWER SEEDS.

Carnatian	Pheaſant-ey'd Pink	Brumpton Stock
Twickenham Stock	Queen Stock	Giant Stock
Bloody Wall	White Wall	Polyanthus
Auricula	Primroſe Tree	Snap Dragon
Roſe Campion	Campanula	Canterbury Bells
Scarlet Lychnis	Columbine	Valerian
Sweet William	French Honeyſuckle	Holyoak
Tree Mallow	Globe Thiſtle	Honeſty or Sattin flow-(er
Everlaſting Peas	Colutea	Bee Larkſpur

March till May; or moſt of them may be ſown in Autumn, in Pots or Boxes of prepared Earth, and defended from the Cold. ——

PEAS.

Early Frame Peas	Dwarf Marrow	White Rouncival
Wrench's Early Hots	Eſſex Reading	Grey Ditto
Early Golden Hots	Nonſuch or Pearl	Roſe or Crown
Charlton Hotſpur	Nonpareil	Tall Crooked Sugar
Ormorod's ditto	Spaniſh Morotto	Dwarf Ditto
Maſter's ditto	Blue Union	Royal Oak
Ledman's Dwarf	Large Marrow	————

October till Jan. for early, Feb. till May for late Sowing.

BEANS.

Barbary, or Maſſagan	Green Genoa	White Bloſſom
Portugal, or Liſbon	Mumford	Broad Spaniſh
Small Spaniſh	Long Pods	Sandwich
Green Nonpareil	Sword or Turkey Long (Pods	Toker, — Windſor

October till Jan. for early, Feb. till June for late Planting.

FRENCH, OR KIDNEY BEANS.

Early Spotted Dwarf	Early Negro	Large White Dutch Runners
Early Yellow ditto	Canterbury Dwarfs	
Early White ditto	Speckled Ditto	Scarlet Runners
Large White ditto	Batterſea Runners	

From March till June and July.

GRASS-SEEDS, &c.

White Dutch Clover	Red Clover	Trefoil	Rib Grafs	St. Foin
	Cow Grafs	Rye Grafs	Burnet	

March till May, or in Auguft or September.

La Lucern,	Crefted Dog's-tail,	French Wheat,	Rape,

March till May, April or May, July.

Riga Flax,	Hemp	Millet	Lentils
Canary,	Maw Seed	Tares	Brown Muftard

March or April.

TREE AND SHRUB SEEDS.

Cyprus Fir	Balm of Gilead Fir	Pineafter
Silver Fir	Larch Fir	New England, or Weymouth Pine
Scotch Fir	Italian or Stone Pine	
Spruce Fir	Clufter Pine	Cedar Lebanon

Feb. till April, or in Septem. or October.

Evergreen Oak Acorns	Beech Maft	Red and white Virginea Cedar
Englifh Oak Acorns	Horn Beam	Pyracantha

Feb. or March. or as foon as ripe.

Spanifh Broom	Bladder Senna	Acacia
Laburnum	Arbutus	———

Feb. till April in good Earth, on a warm Border, or in Pots or Boxes kept warm.

Englifh Broom	French Furze —	China Arbor Vitæ
Candlebury Myrtle	Horfe Chefnuts	Bay Berries
Yew Berries	Evergreen Cytiffus	Laurel ditto
Spanifh Chefnuts	Arbor Vitæ	Feb. till May.

FLOWER ROOTS, *and Time of Planting.*

Ranunculas and Anemonies, ——— September till February.

Hyacinths & Tulips	Ornithogalums	Crown Imperials
Narciffus & Jonquils	Snow Drops	Lilies
Iris's	Crocuffes	—— ——

——————— September till December. ———————

——— Tuberofes, February till May. ———

N. B. Garden Tools, Bafs Mats, Split Peas, Cabbage, Cauliflower and Afparagus Plants, &c. &c. ———

FINIS.

Commentary on the Catalogue

As has been seen already, the catalogue here reproduced in facsimile belongs to the improved type of southern list which, following the London model, arranges the plants under several main heads in alphabetical order of Latin names, with the English names placed opposite. There are four main sections: Hardy Trees and Shrubs; Herbaceous Plants; Greenhouse Plants; and Hothouse and Stove Plants. Added to these, giving common names only, are Fruit Trees; Seeds of Roots, Salads, and Greens with Sweet and Pot-Herb Seeds and Physical Seeds; Tender Annual Flower Seeds; Hardy Annuals; Biennials; Peas, Beans and Kidney Beans; Grass Seeds; Tree and Shrub Seeds; and Flower Roots (bulbs etc.). The extremely perfunctory listing of the last section implies a reliance on imported bulbs from Holland, brought in each year. The contents of each section must be considered in detail, and in comparison with other surviving catalogues of the period.

Hardy Trees and Shrubs

For this section of the list there is a large amount of comparative evidence in the surviving issues of catalogues of other firms of the period, mostly from the Midlands and North of England. From these other lists it is possible to discover the ruling prices for almost all items, and these prices can be confirmed from individual bills. In general, prices were more or less standardized all over the country, and this was due to deliberate price-fixing by agreement between the leading nurserymen. This is known from a letter of 1773 from Perfects of Pontefract, stating that in the previous year 'all the Nurserymen in the South of England . . . came to a Resolution to Advance the Prices . . . and those in the North have done the same'. It was only in the last few years of the century that price-cutting began, with some firms in Scotland and the far North of England, notably William Falla and Son near Gateshead who grew on an immense scale.

The choice of hardy trees and shrubs was very large, though more restricted in regard to forest trees than the northern lists. In Dorset and the neighbouring parts there was in general far less scope for large-scale planting, but on the other hand there are a number of ornamental shrubs which do not appear until much later elsewhere outside London. Thus Galpine was offering three Andromedas: *A. polifolia major* and *minor,* and *A. caliculata* (now known as *Chamaedaphne calyculata*). Though native to Britain, the first of these had only been brought into cultivation in 1768 and even after 1820 plants cost 2*s.* each. *Daphne cneorum*, described as 'most beautiful', had been introduced in 1752 but was still uncommon in English Gardens; about 1800 it was being sold at 4*s.*, a very high price.

To put these prices into perspective it has to be mentioned that common white and Dutch red honeysuckles could be had for 2*d.* each, as could a Sweet Briar or a Weeping Willow. A great number of common shrubs and trees cost only 3*d.*: for example, a Barberry, Box, Bladder Senna, common White or Yellow Jasmine, Laburnum, Laurel, Lilac, Plane tree or Syringa (*Philadelphus coronarius*). Many of the older roses: Blush Provence, Frankfort, Cinnamon, Rosa Mundi or York and Lancaster, cost 4*d.*, and most others 6*d.* or 1*s.* 0*d.*. The Moss Rose might be 2*s.* 0*d.* but was sold at 1*s.* 0*d.* by Brunton & Forbes in Birmingham at the same time, 1782. Very recent roses such as the Burgundy and Rose de Meaux were more costly, for some years standing at about 5*s.* 0*d.* That the Blandford rose was put on the market at 7*s.* 6*d.* shows that in 1789 it was regarded as a great novelty. A few remarkable shrubs and trees of recent introduction were really expensive: Azaleas were 7*s.* 6*d.* each, so was *Calycanthus floridus*, the Carolina Alspice; different species of Magnolia went up to a guinea for *M. acuminata; Rhododendron maximum* cost 15*s.* a plant. What was to become the common (all too common!) rhododendron, *R. ponticum* brought from Gibraltar in 1763 and likewise in Galpine's list, was costing about 4*s.* at the time. The red-flowered Arbutus cost as much.

Galpine seems to have been a pioneer of the heather family, *Ericaceae*, for in addition to three species of Andromeda, two forms of Arbutus and four of Azalea, he stocked three species of Rhododendron including the Alpine Rose Bay, *R. ferrugineum*, and *Erica multiflora* from the Mediterranean. He also had a good selection of conifers: cedar of Lebanon, upright and spreading cypress, the cedar of Goa (*Cupressus lusitanica*), white cedar (now *Chamaecyparis thyoides*, the deciduous cypress (*Taxodium distichum*), both American and Chinese Arbor Vitae, common and 'black' Larch (*Larix x pendula*), silver and Balm of Gilead Fir, Norway, Red and Black Spruce, as well as the 'White' Spruce (now called Hemlock, *Tsuga canadensis*), Swedish Juniper, common and variegated Savin and 'Red Cedar' (*Juniperus virginiana*), Scots Pine, Stone Pine and the Weymouth Pine, and the Yew.

Well over 30 years were to pass before the first introduction of the famous Chinese Wistaria, but Galpine sold its predecessor from Carolina, *Glycine* (now *Wisteria*) *frutescens*, which had been introduced in 1724. He also, rather surprisingly, had another climber from Carolina, *Clematis viorna*, which had arrived in 1730 but has never been a particularly well known member of the genus. From priced catalogues we find that both of these were fairly expensive, at 2*s*. 0*d*. each. On the other hand the American Trumpet Honeysuckle, which had reached the Tradescant garden at Lambeth by 1656, had come down to 1*s*. 0*d*. as had the common Passion-flower, *Passiflora caerulea*. Another valuable hardy climber from North America was *Bignonia* (now *Campsis*) *radicans*, here since 1640 and also priced at 1*s*. 0*d*. in most catalogues. Its relative the Catalpa, however, sold at double the price as it was a relatively recent introduction (1726). The red and yellow flowered Horse Chestnuts, *Aesculus pavia* (1711) and *A. flava* (1764) could be had at 2*s*. 6*d*., but the Fringe or Snow-drop tree (*Chionanthus virginicus*) was 3*s*. 0*d*. The Tulip tree and the Liquidambar or Sweet Gum, as well as the Toothache Tree (now *Zanthoxylum americanum*) were all at half-a-crown (12½p.).

Apart from the strictly ornamental trees, shrubs and climbers, a fair number of ordinary forest trees were stocked, but in the south-west there was evidently no great demand for planting, and the number of varieties was relatively small. There were English, Turkey and Evergreen Oaks, with the American Scarlet and Willow-leaved species and the Cork-tree, but the wide variety of imported kinds of the genus *Quercus*, so much sought after by the landscape planters, were not kept at Blandford. There were the common and purple-leaved Beeches, Ash, common and paper-barked Birch, Hornbeam, Lime, Plane, Sycamore and Sweet Willow as well as the Weeping sort. Only the Poplars, represented by six species, seem to have been popular, and the Elm of all kinds is conspicuous by its total absence. Since there is no suggestion that Dutch Elm disease was prevalent in the eighteenth century, this omission is peculiar and unexplained.

Herbaceous Plants

As has been mentioned already, it is the range of herbaceous plants listed that forms one of the most noteworthy features of Galpine's catalogue. Though these, and many more, could be obtained from the few great nurseries in London, like Malcolm's, it is remarkable that there should have been such demand in Dorset as to warrant this large stock being carried. Much of it could only be justified by large sales of the cheaper items, and we are faced by the problem of what the prices were. A small selection of herbaceous plants are priced in Anderson's Edinburgh list of 1775, but the Scottish prices are undoubtedly far lower than corresponding English ones for the more common plants. For the list as a whole it is necessary to move on some 12 years to Ordoyno's catalogue of *c* 1793-95, from Newark-upon-Trent. This seems to represent fairly closely the general level of costs which prevailed until the inflation of the Revolutionary and Napoleonic wars.

A very few things could be bought for 2*d.* each: Yellow Milfoil, Double Rose Campion (now *Lychnis coronaria*), six sorts of Double Daisies, the Double Dutch Sweet William; but most common plants were at 3*d.*, 4*d.*, or 6*d.* (2½p.). At the other end of the scale were several choice rarities, notably the American *Dodecatheon meadia*, marked by Galpine as 'most beautiful'. This was introduced in 1744, was 2*s.* 6*d.* from Andersons in 1775, later 1*s.* 6*d.*, and by 1800 was down to 1*s.* 0*d.* each. The Perennial Lupin, the Virginian Lungwort (now *Mertensia virginica*), and Rudbeckia (*Erinacea*) purpurea cost from 1*s.* 6*d.* to 2*s.* 6*d.* Other choice plants, costing about 1*s.* 0*d.* each in the 1790s, were the white 'Wolfsbane' or Monkshood (now *Aconitum variegatum* var. *leucanthum*), the Red-hot Poker (now *Kniphofia uvaria*), the double blue Hepatica, *Campanula pyramidalis, Iris susiana,* the double Ragged Robin (*Lychnis flos-cuculi*), a white Phlox, *Saxifraga cotyledon*, what was then known as *S. crassifolia* (now *Bergenia*), the two 'white' Hellebores (*Veratrum album* and *V. nigrum*).

As in trees and shrubs, so in herbaceous plants, variegated leaves were much sought after, to a degree hardly intelligible now. Who, for instance, would pay double the price for a snapdragon (*Antirrhinum majus)* because it was variegated? Another ancient hobby of gardeners was the collection of double-flowered varieties, even of native weeds such as Ragged Robin. In Galpine's list there are doubles of the Rose Campion, the red and blue Hepaticas, Daisies, Marsh Marigolds, several Campanulas, the bloody Wallflower, Lily of the Valley, Sweet William, Sweet Rocket both white and red, *Lychnis chalcedonica*, Ragged Robin, *Lychnis viscaria* (also known as *Viscaria viscosa*), Feverfew, Peony, 'Pilewort' (Lesser Celandine), two other species of Ranunculus, *Saxifraga granulata*, Dropwort, and purple and white Sweet Violets.

In all there are nearly 140 different varieties of plants, and this total does not include bulbs: the tulips, hyacinths, narcissus and jonquils, crocus, iris, lilies, fritillaries, ornithogalums and snowdrops, which would be bought from a separate autumn list for planting from September onwards. We must not overlook the later section of Flower Seeds, which included many annuals and biennials that added greatly to the gaiety of beds and borders. It is evident that colourful flower gardens in great variety were in vogue in the Dorset countryside of 1782. It may be convenient to consider these additional flowers here rather than in their separate place in the catalogue.

Annual Flowers

The seeds of annual flowers are divided, as in most modern catalogues, into two sections for the convenience of the gardener. The Tender (Half-hardy) were to be sown from 'February till May, on a moderate Hot-bed', or 'sown on a warm Border . . . in April, or from March till May'. These included both favourites which are still grown, such as French and African Marigolds and Convolvulus Major and Minor, and a number of plants now almost forgotten or grown by a few specialists for effect. There is also the problem of modern improved varieties, which no longer much resemble the plant as it used to be. Among sorts still in current catalogues are Mignonette ('Minionet' in Galpine), Coxcomb (*Celosia*), *Amaranthus tricolor*, white and yellow annual Chrysanthemum, *Datura stramonium*, Peppers ('Capsicum'), Egg Plant, China Aster, Convolvulus Minor (*C. tricolor*) and Major (*Ipomoea purpurea*), Sweet Sultan (*Centaurea moschata*), Sweet Scabious (*Scabiosa atropurpurea*). Much less familiar nowadays are the Sensitive and Humble Plants, though seed of the latter (*Mimosa pudica*) can still be obtained. 'Diamond Ficoides', better known in modern times as Ice Plant (*Mesembryanthemum crystallinum*) has dropped out of the lists in recent years. The red and yellow 'Zenia Peruvia' (*Zinnia multiflora*), which had reached England as recently as

1770, are now superseded by modern hybrids. The Love Apple, or Tomato, is not generally grown today as an ornament of the flower garden, but it could be. Portugal Bush Bazil was a form of *Ocimum minimum*, and has very likely died out, and the Large Spotted China Pink probably exists only in packets of mixed species and forms. What seem to be totally forgotten are the annual Prussian White, Purple and Scarlet Wallflowers, which seem really to have been Stocks.

Among the hardy annuals very little has altogether disappeared, though in many cases the varieties now obtainable are 'improved' so as to be almost unrecognisable. Forms of *Nolana* still appear in the lists, and so does that old favourite 'Belvedere' or Summer Cypress (*Kochia trichophylla*). Seeds of the Tangier Pea (*Lathyrus tingitanus*) still come in mixtures if not separately, but the Wing Pea (*Lotus tetragonolobus*) is not normally in trade. The curious Horn Flower, Caterpillars, Snail flower and Hedgehogs, species of *Medicago*, have likewise gone out of cultivation. The Ketmia, bladder ketmia or Venice Mallow (*Hibiscus trionum*) is still sown, and *Polygonum capitatum*, an annual still obtainable, is closely related to 'Persicaria'. The Heart's Ease has been marvellously changed into the Pansy, but Virginia Stock, Flos Adonis (*A. aestivalis*), Lavatera and Candytuft remain much what they were. The annual Lupins can still be had in mixture from some seedsmen, though it is nowadays the much larger perennial species and their hybrids that are mainly grown.

Biennials

Because so many of the famous old biennials are commonly bought as plants rather than home grown from seed, their importance as a class is often overlooked. Yet what would a flower garden be if deprived of those standard biennials which were grown two centuries ago, most of them far longer? Pyramid Campanula and Canterbury Bells, Carnation, Columbine, Double Daisy, Hollyhock, Honesty 'or Sattin flower', Everlasting Pea, Polyanthus, Snapdragon, Stocks, Sweet William and Wallflower, even though some of them can be perennial, are best grown as biennials, and form the main body of what Galpine was offering in 1782 under this heading. The Primrose Tree was of course the Evening Primrose (*Oenothera*), still with us; the French Honeysuckle (*Hedysarum coronarium*) is no longer on general sale though it probably can be obtained. Globe Thistle (*Echinops ritro*) and various 'Bee Larkspurs' (*Delphinium*) are still plentiful, and one hopes are likely to remain so.

Greenhouse Plants

From a practical viewpoint, one of the most difficult problems in using Galpine's catalogue, is to sort out from his list for the greenhouse such plants as have in fact proved to be hardy in the open air. The list cuts across the boundaries of woody and herbaceous, fibrous-rooted and bulbous, and requires to be disentangled. All the same, the possessor of a greenhouse still could grow in it most of the 1782 list if he were able to obtain stock. The steep decline of the large greenhouse since 1914 has meant the disappearance from commerce of very many plants which had been grown here from their first introduction until within living memory. On the other hand, the present vogue for experimentation with shrubs, including the marginally hardy species, has preserved on the market quite a few of the entries in Galpine's list.

The rather tender Red-hot Poker is duplicated from the herbaceous section, but *Alstroemeria pelegrina* too can be grown in the open air with some protection. This won Galpine's epithet of 'most beautiful' and, in priced catalogues of the time, cost 2*s*. 6*d*. *Arctotis aspera* which then cost 9*d*. as a greenhouse plant is now represented by hybrids generally sown as half-hardy annuals. Several species of cistus can be grown outdoors in the South, especially near the coast: *C. populifolius* and *C. creticus*, with what is now called *Halimium halimifolium*. Protected conditions in the favoured areas of southern England apply also to Colutea, now *Sutherlandia frutescens, Convolvulus cneorum* which was elsewhere listed at 2*s*. 0*d*., and *Coronilla glauca* (1*s*. 0*d*.). *Erica mediterranea*, then something of a rarity at 3*s*. 0*d*., is now widely grown. Other plants which will succeed in the south, against a wall in a good position or with some slight protection in winter, are *Hypericum balearicum*, the Azores Jasmine, *Lycium afrum, Medicago arborea*, the Bead Tree (*Melia azedarach*) and *Melianthus major*.

The Common Myrtle and some of its varieties, though temperamental, are reasonably hardy in suitable positions, particularly near the sea. *Passerina filiformis* can be grown in the open on the Scilly Isles, but is otherwise a greenhouse curiosity, then and now. It is with the next plant in Galpine's catalogue that we touch the romance of plant history, for 'Philadelphus aromaticus' which then had no English name, is the New Zealand Manuka or Tea-tree (*Leptospermum scoparium*). This was the first of all successful introductions from Australasia, brought over in 1772, and the subject of experimental planting ever since. It dies in really severe English winters, yet will thrive in many places in the South over long periods of average seasons. It was the forerunner of many plants derived from the second British Empire, which in Galpine's time was on the point of replacing the lost American colonies of the first.

Few gardeners now grow the Alexandrian Laurel (*Ruscus hypophyllum*), a tender Butcher's Broom, or *Selago spuria*, but *Teucrium fruticans* and its dwarf relative *T. marum* are worth growing in warm areas. The former, Tree Germander, is one of the few woody labiates that can compare with Rosemary, and is more conspicuous when in flower. When we turn from these half-hardy plants to the strict denizens of the greenhouse, we find that hardly any have stood the test of time. The indoor cyclamen (*C. persicum*) is a conspicuous exception, as well as the hybrid Pelargoniums derived from the species of 'Geranium' listed in 1782. The Oleander too is still very popular, but the vogue for forming immense collections of different species of Mesembryanthemum is a thing of the past. For the rest, only the Winter Cherry, the 'Amomum Plinii' of earlier times (*Solanum pseudocapsicum*), retains through the centuries its singular popularity, now mainly as a Christmas gift.

Hothouse Plants

The inhabitants of the somewhat primitive hothouses of 1782 were a very mixed collection from most of the warmer parts of the known world. In spite of the immense number of later introductions, several of these old plants are still popular, and indeed have taken on a new lease of life as indoor house plants. Such are the Jacobaea Lily (now *Sprekelia formosissima*), which had been grown here for 150 years before Galpine, along with the Atamasco Lily (*Zephyranthes*) and *Haemanthus coccineus*. *Amaryllis belladonna*, although a later arrival, has enjoyed the same permanent vogue. Handsome non-bulbous plants are *Hibiscus rosa-sinensis, Lantana camara*, and the well known Madagascar Periwinkle (*Vinca rosea*). Collections of cactus have also remained popular, along with other succulents such as Crassula and Euphorbia, and the strange Stapelias, of which *S. variegata* is still grown. Galpine's choice of *Chironia* (now *Orphium*) *frutescens* as 'most beautiful' has not been endorsed by continuing cultivation, though the whole genus was still regarded as of great beauty by Agnes Catlow in her *Popular Greenhouse Botany* of 1857. What has always remained at a peak of popularity ever since its introduction in 1754 is the Gardenia. Priced at a guinea (£1.05p) in 1766, it could be had for as little as 7*s.* 6*d.* in 1777, but was still 10*s.* 6*d.* at Newark in 1793-95. Even better known is the 'Cherry Pie' Heliotrope.

Fruit Trees

The Blandford Nursery catered extensively for the table, and stocked many varieties of apples, pears, plums, cherries, and also of the choicer fruits. Very few of these survive in active cultivation now. Among apples there are still the Golden Reinette and Golden Russet; among pears, Catillac ('Cadilliac'), Jargonelle, and the Black Pear of Worcester (or 'Parkinson's Warden'). A greater number of plums still exist as stocks at the National Fruit Trials, but not many are generally propagated except for the Old Green Gage and Reine Claude (the true green gage). The Moor Park apricot, May Duke and Morello Cherries, Royal George peach, and the Burgundy Miller grape are continuing favourites. We cannot hope to sample all the flavours of the dessert that appeared on Blandford tables in 1782.

The Kitchen Garden

Apart from the old types of herbs, hardly any vegetables remain what they were two centuries ago. The pressures of competition and the search for novelty to tempt jaded palates have entirely superseded the famous old varieties well known through the eighteenth and early nineteenth centuries. Probably the Black Spanish radish and the White Cos lettuce are not now very different from what they were, and the less common plants such as salsify, scorzonera and shallots will not have altered much. There were many varieties of the commoner greens, peas and beans, and it can safely be assumed that the customers of Galpine's grounds fared well. What to us is rather surprising is that they seem to have done without the potato.

The Catalogue and its Sources

Galpine's catalogue is most unusual, if not unique, in its title. Instead of a bald statement of the main headings comprised in it, the description includes the additional words 'the most Useful & Ornamental', providing an unexpected human touch. It is, moreover, obvious that considerable trouble was taken over alphabetical order, the correctness of the Latin names, and the correspondence of those in the vernacular. Though closely resembling William Malcolm's large London catalogue of 1778, there are many minor variations that show that Galpine was no slavish copyist. The internal arrangement of species in the larger genera is not alphabetical, and is usually quite different from the order in Malcolm. In the two large lists of greenhouse Geraniums and Mesembryanthemums, however, the order in Malcolm's corresponding series is followed almost exactly. There can be no doubt in these cases that it was from Malcolm that Galpine derived his stock. This is not surprising: most of the provincial nurseries had close links with one or more of the great firms of the London area and could no doubt obtain from them the rarer plants to special order. What is revealed by this country catalogue is not merely a flourishing business at Blandford in the Dorset countryside, but an integral part of the highly organised national trade in plants. Between the scientific explorers on one hand and the professional and amateur gardeners on the other, it was such nurserymen as John Kingston Galpine who transformed the horticultural scene at the midway point between the ancient and modern worlds.

Bibliographical Note

The general history of nurserymen's catalogues is dealt with in John Harvey, *Early Gardening Catalogues* (Phillimore, 1972), and a checklist of surviving lists down to 1850 from firms in Great Britain and Ireland has been issued by the University of Bath Library as *Early Horticultural Catalogues* (1973, with Supplement 1981). Fully detailed descriptions of many book-form catalogues are included in Blanche Henrey, *British Botanical and Horticultural Literature before 1800* (Oxford University Press, 1975), vol. III.

The pioneer work on the history of plant nurseries is Eleanor J. Willson, *James Lee and the Vineyard Nursery, Hammersmith* (Hammersmith Local History Group, 1961), and Miss Willson has now expanded this information in *West London Nursery Gardens* (Fulham and Hammersmith Historical Society, 1982). Both nurserymen and market gardeners are included in Ronald Webber, *The Early Horticulturalists* (David and Charles, 1968), and the earlier development of the trade down to 1800 is covered in John Harvey, *Early Nurserymen* (Phillimore, 1974). A detailed study of two nurseries, based on the rare survival of the firm's archives, is to be found in J. Harvey, 'Two Early Nurseries', in *Journal of the Chester Archaeological Society*, vol. 59, 1976, 66-83.